THE ENERGY LABEL
A MEANS OF
ENERGY CONSERVATION

**Report by the Committee
on Consumer Policy**

The decision to publish the
present document was taken in August 1976

ORGANISATION FOR ECONOMIC CO-OPERATION AND DEVELOPMENT

The Organisation for Economic Co-operation and Development (OECD) was set up under a Convention signed in Paris on 14th December, 1960, which provides that the OECD shall promote policies designed:

— to achieve the highest sustainable economic growth and employment and a rising standard of living in Member countries, while maintaining financial stability, and thus to contribute to the development of the world economy;
— to contribute to sound economic expansion in Member as well as non-member countries in the process of economic development;
— to contribute to the expansion of world trade on a multilateral, non-discriminatory basis in accordance with international obligations.

The Members of OECD are Australia, Austria, Belgium, Canada, Denmark, Finland, France, the Federal Republic of Germany, Greece, Iceland, Ireland, Italy, Japan, Luxembourg, the Netherlands, New Zealand, Norway, Portugal, Spain, Sweden, Switzerland, Turkey, the United Kingdom and the United States.

* *

CONTENTS

INTRODUCTION

1.   For many years growing energy consumption was as self evident
as was economic growth. The availability of relatively cheap energy
contributed to a climate in which energy consumption, costs and resources did not receive particular attention either from industrial
or from private consumers. The recent oil crisis and the subsequent
rises in the price of crude oil and energy in general have entirely
changed this situation and compelled Member countries to concentrate
on energy policy more systematically, taking into consideration all
problems arising in the field of resources, transformation and consumption of energy. In this situation, governments have directed
their efforts towards, on the one hand, the search for new sources
of energy (nuclear, geo-thermal, solar, wind power and tidal energy)
and on the other, towards the development of measures ensuring a
better conservation of the energy available. These latter have the
twin advantage of achieving results both in the short and medium
term and of stimulating technological progress of lasting benefit
regardless of what energy supplies may be developed in the future.

2.   Energy conservation has thus become an issue of major concern for
the relevant authorities in Member countries, and since residential
energy consumption accounts for a considerable proportion of total
primary energy consumption, most Member countries have attempted
to associate the consumer in the fight for energy conservation through
various channels of information.

3.   In connection with energy conservation in the domestic sector,
the question arises whether and to what extent the different means
of consumer information could be employed for the purpose of helping
the consumer economise on his energy costs and whether this would
have a significant impact on the overall energy situation. Among
the various measures aimed at informing the consumer on possible
energy savings - information campaigns, comparative tests, advice
centres - the "Energy Label" for energy consuming household appliances
is the one which is most consistent with the consumer information
concept of conveying essential indications through product related
labelling at the point of sale. Roughly speaking, the aim of this
label would be to clearly indicate to the consumer the efficiency

of a domestic appliance with respect to its energy consumption,
i.e. the ratio between its energy input and its performance, in
order to guide the consumer towards purchasing those products which
make the best possible use of their energy input.

4.   At present, four Member countries, Canada, France, Switzerland
and the United States have introduced, or are planning to introduce,
an "Energy Label" of this kind for certain types of electrical household appliances. In addition, the energy label is under consideration
in Germany, Japan and the Netherlands, and the Council of the European
Communities has issued a Recommendation on the rational use of energy
for electrical household appliances. Since a certain number of other
Member countries have expressed their interest in this subject, the
Committee on Consumer Policy thought it useful to examine the issue
in more detail. Thus, the objective of the present report is to
assess whether an energy label could contribute to energy savings
on the part of private households, and, if so, in what way it could
be formulated and applied. In principle, the energy label could apply
to all types of energy : coal, gas, fuel, or electricity. The present
report, however, deals only with appliances consuming electricity.

5.   In order to better define the problem, the report first briefly
indicates the importance of residential energy consumption and the
main types of household appliances for which energy labelling might
possibly lead to savings in energy consumption of households. The
report then describes the measures taken and the projects which
are under way with regard to energy labelling in the United States,
France, Switzerland and Canada. Finally, the report examines in
detail the objectives and instruments of an energy labelling policy
for domestic appliances and the problems raised.

NATURE AND IMPORTANCE OF RESIDENTIAL ENERGY CONSUMPTION

1. ENERGY CONSERVATION AND THE CONSUMER

6.   The 1974 OECD Report on "Energy Prospects to 1985" (1) devotes
one of its main chapters to the possibilities of energy conservation
in the various fields of consumption : electricity conversion, industry, transportation and the residential/commercial sector. It
distinguishes two types of measures which are expected to result in
energy savings : those which necessitate an initial capital investment and will therefore largely be effective in the long run, and
those measures which can be undertaken in the short term with little
capital investment, e.g. better energy management and increased
energy consciousness on the part of institutions and individuals.
Among the various measures mentioned with regard to the residential/
commercial sector - for instance, energy conscious construction,
improved thermal insulation, district heating schemes using waste
heat from electricity generation or burning municipal waste - the
energy label can certainly be considered one where initial capital
investment would not be too high, particularly when an institutional
framework for labelling and testing consumer products already exists.

7.   However, unlike some other conservation measures which can
either be imposed or directly addressed to a well-defined group of
producers or users of vehicles, premises or energy-consuming appliances,
the actual impact of the energy label as a means of energy conservation depends on a variety of factors and is rather difficult to
determine.

8.   What follows is an attempt, on the one hand, to determine the
areas in which an energy label ought theoretically to be effective,
taking into account the importance of residential energy consumption
as a proportion of total energy consumption and, on the other, to
give some idea of energy consumption represented by the usual range
of domestic appliances.

----

1) Energy Prospects to 1985 - An assessment of Long-Term Energy
   Developments and Related Policies, OECD, Paris, 1974.

9.   With regard to the latter issue, it has to be noted that the
Committee on Consumer Policy is only dealing with the possibility
of savings through a more efficient use of energy, but this is not
meant in any way to express any value judgements on the necessity or
usefulness of certain domestic appliances or on the services they
provide. However, the Committee is fully aware of the fact that the
proliferation of appliances and the trend towards unnecessarily sophis-
ticated items would, in certain product areas at least, justify a
critical assessment of their usefulness both in the interests of
energy conservation and consumer information.

2. EXTENT AND STRUCTURE OF RESIDENTIAL ENERGY CONSUMPTION

10.   The energy statistics available do not allow a detailed ana-
lysis of energy consumption by final utilisation. However, the
above-mentioned OECD report compares the shares in total primary
energy consumption held by the four main sectors - electricity
conversion losses, industry, transportation and residential/commercial
use - indicating the sectoral percentages of total primary energy
consumption in 1972 for the main OECD regions as follows :

SECTORAL PERCENTAGES OF TOTAL PRIMARY ENERGY CONSUMPTION
IN OECD 1972
REGIONAL ENERGY BALANCES (1)

(% of total primary energy)

	North America	OECD Europe	Japan	Australia and New Zealand
Electricity conversion losses	13.8	14.2	12.5	19.0 (2)
Industry (incl. energy sector and non-energy use)	37.3	40.5	52.2	40.6
Transportation	24.2	18.3	15.9	27.0
Residential/Commercial	24.7	27.0	19.4	13.4

1) Energy prospects, vol. II, p. 79.
2) This high figure is due to the predominance of brown coal in
   Australia's electricity generation.

11.   About three quarters of residential/commercial energy consumption is accounted for by space and water heating and air-conditioning.

These utilisations, representing some 15 to 20 per cent of total primary energy consumption, constitute therefore the most important field for energy conservation measures in this sector. Significant savings are in particular expected from improvements in building insulation and temperature control; the hypothetical savings which could be achieved in a new building by improved insulation are some 40 to 50 per cent (1) and metering of heating loads and hot water in apartment buildings and individual billing for energy consumption are supposed to result in energy savings of about 10 per cent. (2)

12. Compared to the energy used for heating and cooling purposes, domestic appliances represent a smaller but still significant proportion of residential/commercial energy consumption. (3) Estimates concerning the percentage of total primary energy consumption represented by domestic appliances are not available. It is however estimated that savings in the residential/commercial sector could be about 20-25 per cent of previously projected consumption in 1985, which, in terms of primary energy, would represent savings of 3 - 6 per cent (4), depending on the country or the region. This is the margin of possible savings to which energy labelling of domestic appliances could, under certain circumstances, contribute.

3. CRITERIA FOR CONSIDERING CONSERVATION MEASURES

13. The growing number and variety of energy consuming household appliances has been a significant feature of post-war economic developments and the trend towards greater comfort by mechanising housework has resulted in a formidable list of energy-consuming household appliances being made available to the consumer.

14. Due to this considerable variety and the relatively frequent introduction of new appliances, it is not even easy to draw up an illustrative list of machines falling into this category. Instead the following rough classification according to the main function of the appliance may help to some extent to define the domestic appliances where energy could be saved :

1. Space heating, cooling, conditioning
2. Water heating
3. Food preservation
4. Food preparation

---

1) Energy prospects, vol. II, page 78.
2) Ibid., page 80.
3) Ibid., page 83.
4) Energy prospects, vol. I, page 85.

5. Household cleaning and maintenance
6. Entertainment and education
7. Lighting
8. Miscellaneous.

15.   Obviously, the relative importance of the appliances falling
under these headings varies widely from country to country, according
to differences in climate, way of life and consumption pattern.
However, though these differences should not be overlooked, the main
aim of the following paragraphs is to give some indication of those
appliances which seem to be more suitable for energy labelling and
those which seem to be less so.

16.   The question of whether and to what extent the energy label or
any other conservation measures in this field can have a significant
impact on total energy consumption has to be examined on the basis
of the following criteria :

   a) the energy consumption of a particular appliance;
   b) the frequency of its use;
   c) its importance on the market.

17.   The figures available in respect of these criteria are in no way
complete, but they should be sufficient for drawing some general
conclusions as to the appliances that could benefit most from energy
labelling.

4. RELATIVE IMPORTANCE OF VARIOUS HOUSEHOLD APPLIANCES ON THE MARKET

18.   The increasing use of energy-consuming appliances has been a
characteristic feature of rising living standards in recent decades,
and ownership of certain appliances has, during the last ten years
for instance, often doubled or tripled. The following figures concerning the development of ownership of three typical products in France
illustrate this trend :

PERCENTAGE OF HOUSEHOLDS EQUIPPED WITH CERTAIN
HOUSEHOLD APPLIANCES IN METROPOLITAN FRANCE (1)

(Percentages)

	1963	1973
TV-set	27.3	79.1
Refrigerator	41.3	86.8
Washing-machine	31.2	65.7

1) Quoted from Annuaire Statistique de la France, 1973 and 1975,
   Institut National de la Statistique et des Etudes Economiques,
   p. 576 and p. 580.

19. As regards the present degree of ownership of the products falling under the previously mentioned classification, it can be assumed that the level of ownership is generally similar in countries with an equal level of economic development. The table given on the next page indicates the ownership per hundred families of a number of major household applicances in the Member countries of the European Communities. (1) Figures for Japan and the United States (2) are given below.

	Japan (August 1975)	United States (1973)
Television sets (black and white)	43.8	69.2
" " (colour)	92.0	55.8
Refrigerators	97.7	80.0
Freezers	-	31.5
Washing machines	97.9	69.0
Dishwashers	-	21.9
Air-conditioner	21.2	48.6

20. These figures show that in some countries, as regards some significant product lines, the majority of households are already equipped with them.

21. The following table provides some more detailed figures with regard to Germany for the year 1973 :

	Ownership per 100 families
Refrigerator	92.5
Vacuum cleaner	90.7
Television set	87.2
Radio set	86.4
Washing machine	74.8
Record player	43.9
Electric sewing machine	37.1
Tumble dryer	32.3
Freezer	28.1
Tape recorder	25.4
Cassette recorder	19.1
Slide projector	19.1
Electric grill	15.8
Ironing machine	9.9
Film projector	7.1
Dishwasher	6.9

1) Quoted from the Harmonized Consumer Survey of the European Communities, Brussels, January, 1975.

2) Source : Statistical Abstract of the United States, 1975; p. 406.

HARMONIZED CONSUMER SURVEY OF THE EUROPEAN COMMUNITIES

Percentage of households possessing durable goods

(In % of total interviewed households)

	Denmark	Germany		France		Ireland	Italy		Nether-lands		Belgium		United Kingdom
	May 1973 1975	May 1973	1975	May 1973	1975	May 1973 1975	May 1973	1975	May 1973	1975	May 1973	1975	May 1973 1975
1. Car	64.9	54.0	53.0	61.9	63.8	60.0	61.1	64.1	59.0	61.1	59.9	59.3	56.0
2. Freezer	39.4	36.7	40.2	8.2	14.6	11.5	-	-	19.0	28.9	20.7	27.5	20.0
3. Refrigerator	79.0	90.3	94.0	85.9	89.0	66.6	91.9	95.3	89.0	92.9	78.1	82.8	85.0
4. Washing machine	56.1	81.5	85.9	61.7	66.0	52.2	74.8	82.0	86.0	88.4	63.9	63.1	72.0
5. Television black/white	70.1	80.7	74.8	73.7	73.0	49.6	88.7	92.2	79.0	67.4	79.9	71.5	40.0
6. Colour television	24.8	16.3	30.8	6.1	12.2	6.3	0.4	1.1	17.0	32.5	6.8	18.3	19.0
7. Dish-washer	12.1	7.7	9.7	4.6	8.1	4.1	11.3	14.7	2.0	7.2	5.8	7.4	3.0

22. The present level of ownership of certain appliances is one important indicator of the possible impact of conservation measures. In addition, the growth rates of sales of some relatively new machines have to be taken into account. A number of appliances are not in widespread use but their sales are increasing at a faster rate than the established appliances. This is, for instance, particularly the case for dishwashers, freezers, colour television sets and stand-by heating appliances. They certainly would have to be taken into account in the context of defining the types of products appropriate for conservation measures.

23. Account must also be taken of the fact that certain appliances are used relatively frequently in one country but are practically unknown in another. For instance, air-conditioners, according to the 1971 statistics, are to be found in 31.8 per cent of United States households and, on the other hand, do not even appear in the European statistics. This may be due to difference in way of life and climate, but it does not necessarily imply that such appliances will never find their way to the European consumer, at least in certain regions.

5. AVERAGE ENERGY CONSUMPTION OF HOUSEHOLD APPLIANCES

24. By way of example, estimates of average energy consumption of certain household appliances in France and the United States are given below.

FRANCE (1)

	Average estimated annual consumption per appliance per year in Kwh	
Refrigerators	300	(estimated for a refrigerator with one door; 2-door appliances are supposed to consume twice as much)
Freezers	900	
Washing machines	450	
Dishwashers	700	(is considered as a very low estimate, could probably be doubled)
Television sets (black and white)	200	
" " (colour)	600	
Electric ranges	1,500	
Electric water heaters	1,500	
Electric and gas cookers	650	
Lighting	250	
Hot running water	1,500	

1) Source : EDF and INSEE.

ANNUAL ENERGY REQUIREMENTS OF ELECTRIC HOUSEHOLD APPLIANCES

Type of appliance	Estimated kilowatt hours consumed
Food preparation :	
Blender	15
Broiler	100
Carving knife	8
Coffee maker	106
Deep fryer	83
Dishwasher	363
Egg cooker	14
Frying pan	186
Hot plate	90
Mixer	13
Oven, microwave	190
Range :	
- with oven	1,175
- with self-cleaning oven	1,205
Roaster	205
Sandwich grill	33
Toaster	39
Trash compactor	50
Waffle iron	22
Waste disposer	30
Food preservation :	
Freezer (15 cu. ft.)	1,195
Frostless	1,761
Refrigerator (12 cu. ft.)	728
Frostless	1,217
Refrigerator/freezer (14 cu. ft.)	1,137
Frostless	1,829
Laundry :	
Clothes dryer	993
Iron (hand)	144
Washing machine :	
Automatic	103
Non-automatic	76
Water heater	4,219
Quick-recovery	4,811
Comfort conditioning :	
Air cleaner	216
Air conditioner, room	1,889
Bed covering	147
Dehumidifier	377
Fan :	
- Attic	291
- Circulating	43
- Rollaway	133
- Window	170
Heater, portable	176
Heating pad	10
Humidifier	136

1) Source : "The Library of Congress, Energy Facts, Novembre 1973",
extract from Statistical Abstract of the United States, 1974.

25. A comparison of the French and United States estimates shows that the figures are fairly similar for each type of appliance. Differences are probably due to different assessments of the frequency of use. The only significant discrepancies occurring with regard to the energy consumption of washing machines and dishwashers may be explained by the fact that the United States models are generally supplied with hot water and therefore consume less electricity.

6. POSSIBLE SCOPE FOR LABELLING

26. The selection from the wide range of appliances available to the consumer of those which, due to their energy consumption, seem to be the most suitable for energy labelling can to a large degree be based upon the frequency of their use. Three categories may be distinguished :

    a) appliances which are permanently in operation;
    b) those which are in frequent and regular use;
    c) those which are infrequently and irregularly used.

27. With this classification in mind, an examination of the illustrative list of appliances and their average consumption (tables p. 13-14) shows that a number of major appliances which would fall within the permanent use category (a) account for the highest proportion of energy consumption : the refrigerator, the freezer and water heating appliances; individual heating appliances and, in certain countries, air conditioners would also fall under this category.

28. In category (b), - i.e. frequent and regular use - the kitchen range would seem to be the biggest consumer of energy. Compared to this, other appliances such as washing machines and dishwashers are less important, though they still consume a considerable amount of energy.

29. In category (c) - infrequent or irregular use - appliances with a high energy consumption are relatively rare. Either their operating consumption of energy is very low or they are so infrequently used that their consumption does not have any significant effect on total annual consumption.

30. It can therefore be concluded that energy labelling could be quite effective if it concentrated on those major household appliances which are in permanent or regular use.

MEASURES TAKEN BY MEMBER COUNTRIES WITH A VIEW TO THE ADOPTION OF AN ENERGY LABEL FOR DOMESTIC APPLIANCES

1. GOVERNMENT VIEWS, PROGRAMMES AND PRESENT STATE OF WORK

31. As already indicated in the introduction, Member countries do not attach the same importance to the creation of an energy label. Some countries have taken no action in this field and do not expect to take any in the near future; others recognise the importance of giving the consumer more information about potential energy savings, but are still considering what type of measures might best be taken to this end.

32. However, all these countries are interested in receiving information on the progress of current experiences in this field.

33. Up to now, only four Member countries : Canada, France, Switzerland and the United States, have formulated schemes for allocating an energy label to energy consuming household appliances. (1) In addition, energy labelling is under consideration in Germany, Japan and the Netherlands and has been studied by the Commission of the European Communities, whose Council issued a Recommendation on the rational use of energy for electrical household appliances on 4th May 1976. (2) The United States and Canada have recently decided to operate energy labelling on a statutory basis, whereas in the other cases mentioned it is applied or envisaged on a voluntary basis. All the schemes are set out to cover appliances which consume a substantial amount of energy and are sold in large numbers, e.g. washing machines, dishwashers, refrigerators, freezers, heaters and cooking appliances. In Sweden, the subject of energy labelling is part of a wider project comprising in particular the question whether it is possible to initiate production of household appliances consuming less energy than those produced now.

34. The objectives of establishing an energy label are fairly similar in all the projects mentioned above and may be summarised thus :

---

1) The products to be covered in each of the four Member countries are listed in the comparative table given in Annex I.

2) O.J. No. L 140 of 24th May 1976, p. 18.

a) to provide consumers with information presented in a uniform, easily accessible manner, to enable them to compare the energy consumption of domestic appliances sold for a specific purpose and make their choice accordingly;

b) to stimulate advances in the design of domestic appliances with a view to greater energy saving.

35. In the statements preceding practical work on these projects, the prospects of achieving these aims were judged very positively. Announcing the United States (voluntary) scheme in October 1973, the spokesman for the Department of Commerce noted that a saving of only 5 per cent on household energy consumption was equivalent to the consumption of 300,000 households. In France, it was expected that the scheme should lead to an energy saving of between 5 and 10 per cent for the appliances concerned, without, of course, affecting the services rendered.

36. The energy label was created in the United States following directives given by the President in his 1973 energy message. In France, preparatory work on that project started in early 1974 as part of the government's general policy of encouraging energy savings.

37. In fact, the energy label is at present in operation on a voluntary basis in the United States and in France, where the first energy labels have appeared on household appliances during the year 1976, and is in the planning stage in Canada. Since at present there is no additional information available with regard to the changeover from voluntary to mandatory labelling in Canada and the United States, the following Sections 2-5 refer, as regards those two countries, to the concepts and realisations developed under the voluntary schemes.

2. COMPETENT AUTHORITIES

38. In Canada, the planning and implementation of the energy label, the so-called "CANTAG" programme, come within the competence of the Ministry of Consumer and Corporate Affairs, in co-operation with the Ministry for Energy, Mines and Resources, the national standards organisations, and representatives of industry, commerce and consumer interests.

39. In the United States, the Federal Energy Administration has overall responsibility for energy labelling. It directs the National Bureau of Standards (part of the Department of Commerce) to develop test procedures to indicate energy consumption and operating cost. The Federal Trade Commission has enforcement responsibilities. There are close links between the Federal and State bodies and local

authorities. State governments and local authorities are being urged to promote the Federal Programme rather than try to implement their own energy label schemes.

40. In France, the initial steps were taken by the Energy Saving Agency which is also managing the programme, whereas the French Standards Association (Association française de normalisation - AFNOR) is in charge of defining the test methods. This work is jointly undertaken by representatives of the production sectors concerned as well as of distributors and consumers. The form and layout of the energy label as such is chiefly developed within the framework of the French Association for Informative Labelling (AFEI - Association française de l'étiquetage informatif) in accordance with a voluntary procedure. The undertakings given by manufacturers to introduce voluntary labelling should result in the use of mandatory requirements in only exceptional cases. (1)

41. In Switzerland, consumer organisations have been striving for several years to reach agreements with the various branches of manufacturers of electrical appliances on a systematic informative labelling scheme. With regard to electrical household appliances, these efforts have lead to the creation of a uniform label indicating the main characteristics which are of importance for the consumer in connection with a largely distributed brand. Later on, consumer organisations, together with the Swiss Institute for Refrigeration Systems, have developed standards for the labelling of freezers, refrigerating cupboards and refrigerators. In co-operation with the associations of manufacturers of washing machines and of electrical appliances, the consumer organisations have recently succeeded in establishing guidelines concerning energy labelling for electrical household appliances of all kinds. From the legal point of view, these guidelines are recommendations addressed by the above organisations to manufacturers and distributors. According to these recommendations, the label has to include the important features of a product and indicate its characteristics of use in an objective and uniform way, thus allowing them to be compared and controlled. Energy consumption has to be indicated on this label in so far as it can be correctly calculated. This system is already used by various producers and is likely to be more widely applied in the near future.

---

1) There is one exception to the principle of voluntary labelling which concerns the petrol consumption of automobiles sold in France. From 1st April, 1976 on, any mention of consumption in any advertising or any trade material will have to refer to the indication of this consumption in accordance with a standardized method of measuring determined by a ministerial order (Order of 21st April, 1975).

42. In Japan, the Ministry of International Trade and Industry
is in charge of developing a voluntary energy labelling scheme, and
in the Netherlands the Ministry of Economic Affairs is the competent
authority in this field.

3. MEASURING AND INDICATION OF THE OPERATING COSTS
OF APPLIANCES AND THEIR ENERGY EFFICIENCY

43. In the United States, the measuring of the performance of an
appliance in relation to its energy consumption and the use made
of the results of such tests varies according to whether it is possible to determine an average consumption under normal (standardized)
conditions for use for a given period, a month for instance, or
whether it is possible to express the relationship between energy
input and technically measurable performance in one abstract figure,
i.e. the energy efficiency ratio. Thus, the applications of the
energy label in the United States fall into two categories.

44. In the first category, for example, for a refrigerator/freezer,
the label shows :

  i) the particulars of the appliance (manufacturer's name and
     model number, features);
 ii) the energy consumption (expressed as kWh per month) under
     standard test conditions;
iii) the corresponding monthly cost (for average use) worked out
     in dollars on the basis of local electricity tariffs. The
     cost of defrosting the appliance is also given in dollars;
     this cost varies according to whether the appliance is fitted
     with an automatic or semi-automatic defroster.

45. In the second category, for an air-conditioner, for example,
it is not possible to work out the monthly running cost for normal
use because of the great differences in climate and conditions of
use (living-room or bedroom) to which this equipment is exposed. In
this case, the label basically gives the following information :

  i) the particulars of the appliance;
 ii) the cooling capacity, calculated in British Thermal Units
     (BTU) per hour of operation. The British Thermal Unit is
     equivalent to 252 calories and is the amount of energy pro-
     duced by the combustion of an ordinary wooden match;
iii) its energy efficiency ratio (EER), obtained by dividing its
     cooling capacity in BTU by the energy used in kWh. The appli-
     ance with the best energy efficiency is the one with the
     highest EER.

46.   In <u>Canada</u>, a similar method of testing and information is
planned, including in particular the preference for expressing the
average consumption in terms of money.

47.   In <u>France</u>, the energy label is part of a more general labelling
scheme developed by the AFEI. It contains :

  a) <u>On the front</u> :

    i) the general characteristics of the appliance necessary
       for the information of the consumer;
   ii) its energy consumption for a standardized performance.

  b) <u>On the back</u> :

    recommendations relative to the installation and use of the
    appliance, in particular from the point of view of energy
    conservation.

48.   In <u>Switzerland</u>, energy labelling is part of a more comprehensive
labelling scheme for electrical household appliances, indicating
all the important characteristics of use for a certain product. Models
for such labels, which are available from France, the United States
and Switzerland are shwon in the attached Annex. While the United
States label centres on conveying the energy cost, the French label
gives the energy consumption among a number of other indications.

4. PROCEDURES FOR OPERATING, AMENDING OR WITHDRAWING LABELS

49.   The procedure set up under a voluntary programme in the United
States by the Department of Commerce is highly elaborate and calls
for detailed examination. A preliminary draft relative to this procedure was published in the Federal Register on 5th June, 1973. The
final version, incorporating views expressed at the public enquiry
into the initial draft, was published in the Federal Register on 26th
October, 1973. These procedures will remain in effect until supple-
mented by those produced under the Energy Policy and Conservation
Act (Pl : 94-163) of 22nd December, 1976.

a) <u>Content of specifications</u>

50.   The text of October, 1973, gives a detailed description of the
content of the specification for a given type of appliance :

    i) description of this type of appliance and the relevant range
       of energy efficiencies;
   ii) test methods used for determining the energy consumption
       or energy efficiency of appliances of this type;
  iii) specimen of label and instructions for affixing for easy readin

iv) participation of manufacturers in the programme.

b) Test methods

51.   The test methods are those defined in the standards. If these
do not exist, ad hoc standards may be formulated by the Department
of Commerce in co-operation with the other parties concerned.

c) Publication of specifications

52.   Any interested person may submit comments in writing within
30 days after the publication of a draft specification in the Federal
Register. Persons wishing to add to these comments by expressing
their views at a hearing may do so within 15 days following the
publication of the draft specification in the Federal Register. All
written and verbal comments will be printed in the Federal Register.
On expiry of the period of 30 days, the Department of Commerce will
publish in the Federal Register a notice comprising :

   i) either the final version of the specification, inviting
      manufacturers to submit their requests to participate;
  ii) or an announcement that a further period will elapse before
      the final text of the specification is published;
 iii) or an announcement that the draft is being withdrawn for
      further examination.

d) Participation of manufacturers

53.   In expressing their wish to participate in the programme, manufacturers agree to adhere to all the provisions set out in the specification. Within 30 days of submitting his request the manufacturer
must provide the Department of Commerce with a description of his
test methods; he is bound to preserve the results of this test method
for two years after production of the model concerned has been ter-
minated. The manufacturer undertakes to comply with any reasonable
request made by the Secretary for Commerce that he carry out quality
control tests on his appliances, to ensure that they comply with
the provision of the relevant specification.

e) Amendment and withdrawal

54.   The specification may be amended at the request of the Department of Commerce using a procedure identical to that used in its
initial preparation. If the Department considers that a manufacturer
is not honouring his commitments, it may decide to terminate the
manufacturer's participation. The manufacturer is allowed 30 days
to show cause why this should not be done. A manufacturer may at
any time withdraw from the programme by giving written notice to the
Department.

f) Annual report

55.   The Department of Commerce will publish an annual report on
the progress of the programme; this report will list the manufacturers
participating in the programme and the models included in it.

5. CONSUMER INFORMATION AND CONSUMER RESPONSE TO THE LABEL

56.   Although a uniform, conspicuous and attractive appearance of
the label is a basic condition for its informative value to the consumer, the countries applying or planning them also recognise the
necessity to amplify the information provided by the label as such
by various publicity measures. Thus, in the United States, manufacturers may reproduce the label in their advertising material, on
condition that the whole label is shown. The bulletin published by
the Office of Consumer Affairs of the Department of Health, Education
and Welfare carries periodic articles about progress with the energy
label. Brochures describing the applications of the label to various
kinds of appliance are published by the Department of Commerce National Bureau of Standards, and are widely available where appliances
are sold. These brochures give a clear and practical illustration of
the information the consumer can obtain from the label. In France,
the Energy Saving Agency organised a consumer information campaign
when these labels were first allocated, not only locally at sales
points but also through the press, radio and television.

57.   Due to the very recent introduction of the label, any general
assessment of its actual impact on the consumer's buying decisions
and on product planning would be premature. The United States , conveying the only experience available at the moment, stated that the
introduction of the energy label for air-conditioners had caused
a considerable shift towards the less energy consuming models.

6. COMPULSORY ENERGY LABELLING IN THE UNITED STATES

58.   In the United States, under the Energy Policy and Conservation
Act (PL 94-163) of 22nd December, 1975, energy efficiency labelling
became mandatory for all major appliances. Three Federal agencies
were given responsibility and authority under the law for developing
the new labels. The Federal Energy Administration (FEA) is in charge
of implementing the law and will work with other agencies to get
information needed to design the labels. The Federal Trade Commission
(FTC) has the authority to actually create the labels and to specify
what information is to be used and how. The National Bureau of
Standards (NBS) is charged with developing test procedures that FEA

can adopt for use by industry to arrive at the information required on the label. Under the law, NBS must develop and propose test procedures by June 1976 for refrigerators, refrigerator-freezers, freezers, dish-washers, clothes dryers, water heaters and room air-conditioners. Within 30 days the FEA must modify or elect some other procedures, following which the FTC must, within 30 days, publish a proposed labelling rule. FTC then has 60 days to review and publish a final labelling rule which will have the force of law. By September 1976, text procedures must be developed for heating equipment (not including furnaces), television sets and kitchen ranges; by June 1977 test methods must be proposed for clothes washers, humidifiers, dehumidifiers, central air-conditioning and furnaces. While this process under law proceeds, the Department of Commerce will continue to maintain and service the four specifications which have to date been issued in final form under the Voluntary Labelling Programme for Household Appliances and Equipment to Effect Energy Conservation. These four specifications cover room air-conditioners, refrigerators, combination refrigerators-freezers and freezers.

7. ACTIVITIES AT EUROPEAN LEVEL

59. In its "First periodical report on the Community action programme for the rational use of energy" which deals with the realisation of Community objectives in this field, the Commission presented, among other proposals, a Draft Council Recommendation on the rational use of energy with regard to the consumption of electrical household appliances. This draft was adopted by the Council of Ministers on 4th May 1976. (1)

60. This document is a Recommendation of the Council to Member states to :

"adopt any measures necessary to ensure that :

a) the unit energy consumption of each electrical household appliance listed in the Annex (2) hereto is indicated on a label in conformity with harmonized European standards for the information of prospective buyers. The main purpose of these harmonized standards would be to define a common method of labelling and of informing consumers about the energy consumption of electrical household appliances, and

---

1) O.J. N° L 140 of 24th May 1976, p. 18.
2) The annexed list of electrical household appliances contains water heaters, cookers, refrigerators, freezers and deep freeze units, television sets, dishwashers, washing machines, dryers, spindryers, ironing machines.

to define a method of measuring the unit consumption, compliance with these standards would be ensured in accordance with procedures which do not impede free movement of goods within the Community;

b) the same indications of unit energy consumption are used both in consumer information and in advertising, thus providing the consumer with comparable energy consumption figures on which to base his choice;

c) an information campaign is undertaken in each country to make consumers aware of the way in which each of the electrical household appliances listed in the Annex hereto should be used in order to achieve maximum energy-saving.

III

OBJECTIVES AND INSTRUMENTS OF AN ENERGY-LABELLING
POLICY FOR DOMESTIC APPLIANCES

1. A PARADOXICAL OMISSION IN THE
INFORMATION GIVEN TO CONSUMERS

61. As pointed out in the first Chapter of this report, the number
of domestic appliances used in Member countries has been growing
rapidly and in a few years big markets have opened up for new consumer durables. This development has taken place until recently
in a steadily expanding economy with high, and sometimes very high,
growth rates and relatively low energy prices, and it has been
encouraged by the spread of credit selling.

62. The technical features of the appliances manufactured during
this period reflect this state of affairs and competition between
manufacturers has tended to take the following two forms.

a) Manufacturers have aimed at low selling prices in order
to capture a clientele more concerned about what an appliance costs to buy than about what it costs to run throughout its life. As a result, there has been a price-cutting
race in selling certain consumer durables at the expense
of their energy efficiency. Thus, in some ten years, maximum
energy consumption of refrigerators allowed under French
regulations has been raised substantially in order to take
account of competition in the form of price cuts and lower
energy efficiency which developed for these appliances in
the Common Market. Over this same period, a similar trend
was seen in other sectors , including the building sector.

b) At the same time, manufacturers have tried to persuade
households already owning consumer durables to buy new
equipment by offering them appliances featuring technical
improvements which could step up energy consumption without
necessarily affording equivalent gains in usefulness.

This twofold policy on the part of manufacturers has led by different paths to the same result, namely higher energy consumption by the appliances.

63. While notable progress was made in several sectors (including food products, textiles and cosmetics) in giving consumers information by labelling, the labelling of domestic appliances still suffered from a serious omission in that consumers were not told how much energy was required to operate them. Thus essential information was lacking for assessing the quality-price ratio of the appliances on the market and, as the cost of energy rose, its omission became still more damaging to the interests of households and the economy. Indeed, an appliance which has to operate continuously (like a refrigerator-freezer) or every day (like a dish-washer, air conditioner or television receiver) may consume during its life an amount of energy costing as much as its purchase price or more, and the gain to the consumer who chooses an appliance with a higher energy efficiency than competing appliances may often be considerably greater than the extra purchase price involved. (1)

64. The value of energy labelling has not claimed the attention of all Member countries to the same extent and this is to be explained by a number of reasons, including the following :

a) the apparent absence of any real interest on the part of the majority of consumers whose attention, instead, is attracted by other criteria in the use of domestic appliances;

b) the difficulties (to judge by a number of recent experiments made by certain Member countries in this field) inherent in any kind of informative labelling and those involved in comparative energy efficiency testing and in expressing the findings in a way that is both precise and immediately understendable to the consumer;

c) the constraints which even voluntary energy labelling inevitably imposes on manufacturers and the effects that it may have on the free play of market forces;

_____

1) This is illustrated by the information given in a consumer folder issued by the United States Department of Commerce which draws attention to the relationship between the purchase price of an air-conditioner and what it costs to run (a function of its energy efficiency ratio - EER). Model A, with a higher EER (9.3), costs $ 60 more to buy than Model B, whose EER is only 5.8. The model with the 9.3 EER produces a saving of $ 21.84 in energy consumption every year compared with the other model so that the difference in purchase price is worked off within three years.

	Model A	Model B
Hourly cooling capacity (in BTU)	8,000	8,000
Consumption (in watts)	860	1,380
Energy efficiency ratio (EER)	9.3	5.8
Hours in use each year	1,400	1,400
Cost of energy consumed each year	$ 36.12	$ 57.96

d) more generally, the possibility that governments may react to higher energy prices by looking for new sources of energy rather than by introducing measures designed to conserve the energy that is available;

e) in some Member countries, the price of electric energy is still low and the possible savings by means of an energy label are considered to be too small to justify work on an energy label.

2. PURPOSE OF ENERGY-LABELLING

65. Energy-labelling has three purposes :

a) Initially, to enable the consumer to make a reasoned choice between the products offered to him and thus to reduce energy consumption, i.e. day-to-day expenses.

b) In the medium term, to provide an incentive to manufacturers to design appliances with improved energy performance, and therefore better tailored to consumers' requirements. This could well lead to a "new generation" of appliances matching the new situation and the new relationship between purchase price and running costs, although the changeover can only be gradual for the following reasons :

   i) manufacturers must find it compatible with the depreciation of their existing equipment; with the cost of modifying the production process, and the time this takes, and lastly the disposal of any stock;

   ii) as far as consumers are concerned, the changeover must allow for the time that is necessary for their habits and behaviour to change, which implies a relatively long phase of demonstration, information and education.

c) Ultimately, in the economic field, to help national economies towards satisfactory equilibrium in their energy balance and their balance of payments, and, in the ecological field, to promote optimum conservation of non-renewable energy resources whilst encouraging environmental protection.

66. Such being the case, it is doubly desirable for this rather long process to be started without delay. It will benefit all the parties concerned and governments may guide and support the measures taken by the trade to this end by providing testing facilities, reference material, and arrangements for supervision and publicity.

67. In conclusion, energy-labelling applied to all domestic appliances is clearly one way of opening up a new avenue to technical progress

in an expanding sector of industry. It will promise success to the
more dynamic enterprises in competing with their rivals and will
be a useful addition to national economic policy, an example being
the United States, where in the first few months of trying it out
consumers showed their preference on technical grounds for high
efficiency air conditioners.

3. NATURE AND CHARACTERISTICS OF AN ENERGY-LABELLING SCHEME

68.   Although many Member countries are considering energy-labelling,
it has only been applied in very few cases so far. However, these
cases and the experiments which led to them bring out the nature
and main characteristics of an energy-labelling system. They are
summed up in the four following points :

   a) Drawing up a programme on a joint basis specifying the
      categories and models of appliance to be labelled with their
      energy consumption;
   b) Prohibiting all manufacturers and dealers from giving the
      public any information on the energy consumption of the
      appliances in the scheme that is not in conformity with
      the results of a standardized and officially approved
      energy performance test. One cannot allow two methods to
      be used in the market for informing consumers about energy
      consumption, one of which is rigorous, accurate and autho-
      rised, while the other is left to the free choice of the
      advertiser. To do so would create unfair conditions of
      competition between manufacturers and mislead consumers.
   c) Devising a reproducible performance-testing technique. The
      test method is finalised under the control and supervision
      of the public authorities, after consulting both buyers
      and sellers and with their co-operation.
   d) Defining a standard method of indicating the performance
      thus measured, i.e. a system of labelling appliances to
      serve as a reference for all publicity material and trade
      literature about them.

The system of labelling may either be voluntary, i.e. open to manufacturers who wish to join it and who comply with the specifications
laid down, or be made legally compulsory for manufacturers of certain
categories of appliance.

69.   Both labelling systems, voluntary and compulsory, are in use
in various Member countries.

a) The former is being used in France and there are other
   countries, including Germany, which prefer the voluntary
   principle. There is reason to believe that it may also be
   adopted by the Member States of the European Communities
   where an energy-labelling scheme for domestic appliances
   is being studied. The voluntary system is in line with the
   competitive aspect of improving the energy efficiency of
   appliances. It maintains complete freedon of expression
   for consumer wishes and manufacturers' initiative. It may,
   however, meet with difficulties and prompt certain reser-
   vations as is clear from what has happened in recent experiments conducted in the United States and Canada.

b) In the United States, voluntary labelling (introduced in
   1973) was replaced by compulsory labelling in December 1975.
   Canada is following suit; originally in favour of the voluntary system, it is now working on a compulsory labelling
   system after coming to the conclusion that voluntary labelling
   would cover far too small a share (about ten per cent) of
   the domestic appliance market. It seems that manufacturers
   themselves, realising the need for balanced conditions of
   competition, regard compulsory labelling as likely to be
   more fair and more effective.

c) The choice between the two systems is not, therefore, irre-
   versible. They may even, to some extent, be combined, in that :

   i) A voluntary labelling system could include some degree of
      government support to firms joining the scheme. This
      support may include technical assistance and a substantial
      contribution to national publicity for the scheme.

   ii) A voluntary labelling system could well, in time, lead
       to legally compulsory standardization whereby a maximum
       consumption would be laid down for certain categories of
       appliance and the labels on them would have to indicate
       the legal standard. It is true that a legislative control
       has more chance of being effective when it is based on
       actual experience than when introduced to the market
       without any prior study.

4. ENERGY-LABELLING AS ONE OF THE INSTRUMENTS OF ENERGY CONSERVATION POLICY

70. Choosing an energy policy is a policy decision on which the
technical measures taken to implement it depend, so that energy-
labelling involves two distinct stages, a policy stage followed by

a technical stage. In the "policy" stage, there are various possible aspects that do not necessarily all arise, or arise in the chronological order given below which is purely illustrative. They are, in fact, various aspects of the same activity.

a) First, public opinion must be informed by a high-level source of the aims of a national policy for conserving energy in the interests of consumers as well as those of the general economy and every citizen must be invited to take part in it. An example is the "energy message" delivered by the President of the United States in 1974.

b) Specialised government committees may be set up to carry out surveys and hold consultations regarding this problem. Th~ ould include representatives of the responsible autho-
.~ties, manufacturers, dealers and consumers and would be assisted by qualified experts. (1)
Their tasks would be :

   i) To conduct a country-wide survey and hold wide-ranging discussions on the objectives, conditions and instruments of a policy for conserving energy , including energy-labelling;

  ii) To report to the government on their findings and to propose a plan of action;

 iii) To guide research work by government and private enterprise and make public opinion aware of the implications of the proposed action (see paragraph 65 above) for :

   - the household budget;
   - the health of the national economy;
   - the wise use of non-renewable resources;
   - the protection of the environment.

c) While these special committees are so engaged, the permanent advisory bodies attached to the central government or local authorities might be invited to study the problem and give their opinion. In this way, all groups concerned with social and economic policy would be given a say in preparing and applying the proposed labelling system.

d) Consumers should be informed of the advantages of energy-labelling, and the progress it represents, and should receive, via the mass media, the necessary informative and educative material enabling them to understand the objects of the scheme and to derive maximum benefits from it.

---

1) Examples in France are the "Comité consultatif de l'utilisation de l'énergie" (Advisory Committee on the Use of Energy) and the Working Party on Waste Prevention.

e) It would then be for the government to decide which department(s) would have special responsibility for conducting operations, in conjunction with the standards authorities and non-governmental organisations. (1)

5. TECHNICAL STAGES IN INTRODUCING ENERGY-LABELLING

71. Five main sectors are involved in energy-labelling, namely the authorities, the enterprises which distribute energy, appliance manufacturers, dealers, and consumers. All need to co-operate continuously throughout the seven stages listed below, the order of which has no significance because, in fact, they are interlinked and concurrent.

a) Development and formulation of standard performance-testing methods for appliances.

b) Indicating the results of these methods by means of a labelling system.

c) Definition of the field to which the labelling system is to apply.

d) Assessment of the cost of labelling to the producer.

e) Financing the labelling system.

f) Procedures for assigning labels, supervision and imposing penalties.

g) Publicity for the labelling scheme.

Each of these stages calls for comment :

a) Development and formulation of standard performance-testing methods for appliances

72. The problem is to devise a reproducible technique for determining the quantity of energy consumed by an appliance in producing a certain result. The definition of this result will refer to a number of parameters such as :

- for a refrigerator, the extent to which it is filled, ambient temperature, door-opening frequency, defrosting frequency, etc., or
- for a dish-washing machine : type of dirt on the plates, water quality, water temperature, washing cycle, type and concentration of detergent, criteria for judging what constitutes "clean" cutlery or glass.

---

1) In the United States, there are the "Federal Energy Administration" and the "U.S. Energy Resources and Development Administration" and in France the "Energy Saving Agency", attached to the Ministry of Industry and Research.

73. The ratio found in this way between energy consumed and result obtained can be expressed in various ways.

74. The first method is to determine the consumption of an appliance on a standard basis reproducing the assumed conditions of its average real use. This is relatively simple to arrive at for an appliance which operates continuously, e.g. a refrigerator. All that has to be done is to work out a pattern of use for the appliance, approximating to normal service, and to find the quantity of energy consumed in running the appliance for a specific period of time. This is the energy-labelling method used in the United States and in France for combined refrigerator-freezers.

75. The second method consists in evaluating the energy efficiency of an appliance in terms of a ratio expressing the quantity of energy necessary to obtain a result measured in constant units. This method is suitable for testing the performance of appliances with greatly varying conditions and times of use. It is used in the United States for a number of appliances including air conditioners. An appliance's energy efficiency ratio (EER) is obtained by measuring its output in British thermal units (BTU) for one hour of service and by dividing this figure by the number of kWh consumed. The higher the EER, the higher the energy efficiency of the appliance. In the United States, an air conditioner's energy label, apart from showing its energy ratio (e.g. EER : 9.3), also indicates the maximum and minimum ratios available on the market (in this case, 9.9 and 5.4) so that a would-be purchaser can immediately see how any one air conditioner compares for efficiency with the competition.

76. It is obviously highly important that performance-testing methods, on which the information given on the appliance label is based, be standardized and identical for appliances sold on the same national and international market. Otherwise, energy performance figures would be meaningless to the consumer (since he would be unable to make a fair comparison between two appliances) and would distort competition.

b) Indicating the results of performances-testing methods by means of a labelling system

77. The first need is for the consumer to be able to see, read and understand the label at once. Showing the EER is an easy solution; the highest figure identifies the appliance with the highest efficiency. In the absence of a ratio, efficiency can be shown on a performance scale, the lowest point of which (on the left hand side) corresponds to the minimum performance required by the standard (or failing this, lowest market performance) and its highest point (right

hand side) to highest market performance. The efficiency of the model concerned is entered on the performance scale (which rises from left to right) and the index indicating this efficiency some-where between the two extremes shows its relative value. This is the system used in France (see model of label in Annex II).

78. A small margin of error is accepted in the indication given on the performance index label in order to allow for the scatter that is possible at both manufacturing and testing levels. The standard should lay down the percentage tolerance; if it is exceeded, the right to use the label should be reconsidered.

79. The question arises whether it is better for an appliance's energy consumption to be shown on one label giving all the infor-mation that the purchaser ought to have (technical data, advice as regards purchase, installation and use, coverage of guarantee and details of after-sales service) or whether it is better for a separate label to be used. The United States and France have answered this question in different ways, France opting for the first solu-tion and the United States for the second. Separate labelling is certainly a better way of spotlighting the aspect to be stressed, namely energy efficiency, but it has the drawback of adding a further label to those already in existence. Combined labelling offers the advantage of presenting the consumer with a single item giving him all the information he needs about the product concerned, in which case it may be thought that the single document would stand a better chance of being read carefully and kept. This system also avoids the extra cost and handling caused by additional labelling. Germany and Belgium have indicated their preference for combined labelling.

80. The cash value of an appliance's consumption may also be shown on the label. In the United States, for example, the label on refri-gerator-freezers shows the cost of one month's operation of the appliances, with a number of variants for the regions where it may be used, allowing for the different electricity charges in them. Electricity tariffs and market conditions are fairly uniform in Canada, as they are in the United States, so that showing the cash value of an appliance's energy consumption on the label is perfectly feasible; elsewhere it might be difficult to do, as in France where electricity tariffs vary with locality, time of day and consumption bracket. In such cases, however, the information a user needs in order to be able to work out what the kWh ratings on the label mean in terms of cash can be given in literature provided at the point of sale or by the various offices qualified to give out such infor-mation, i.e. electricity undertakings, standards or labelling orga-nisations, government offices and consumer organisations. Lists of

these sources of information are generally printed on the literature provided for customers at the point of sale.

81. There is everything to be gained by giving the energy label an identifying mark in the form of an easily recognisable symbol which would be the same for all appliances and would be reproduced on all labels and literature. The symbol could be used in conjunction with the national mark authenticating the label in the eyes of the purchaser. Examples are the "CANTAG" mark in Canada, the national mark issued by the United States Department of Commerce, and - in France - the "AFEI" (Association française pour l'étiquetage d'information - French Association for Information Labelling) label and the national NF (Normes françaises) mark for French standards.

c) Definition of the field to which the labelling system is to apply

82. The choice of appliances to be included in an energy-labelling scheme depends on a number of criteria :

   a) level of energy consumption of the appliance;
   b) frequency of use of the appliance;
   c) number of appliances in use on the market.

These three criteria determine the share of household budgets that the appliances account for. Other considerations in the preparation of a scheme are new products which may be heavy energy consumers and appliances incorporating technological advances that increase their energy efficiency.

83. In a voluntary labelling scheme it is important that a high percentage of the appliances available on the market should be covered in each category. The target, to be aimed at, on a gradual basis, should be 80 - 90 per cent of the number in use. Whether the labelling scheme is voluntary or compulsory, the total number of appliances covered should account for more than half total domestic energy consumption. This will make it possible to cover the cost of running the scheme and to ensure its full effectiveness for consumers.

d) Assessment of the cost of labelling to the manufacturers

84. Energy labelling saddles manufacturers with two kinds of cost - direct and indirect.

   a) Direct costs are mainly those involved in installing energy performance-testing systems and in testing appliances for compliance. They vary considerably with product and certification requirements. They may be relatively low in cases where appliances already satisfy the requirements laid down

in the standards and very much higher when they do not. They
also depend upon the degree of accuracy with which consumption
performance has to be determined and the strictness of the
tolerances allowed (admissible deviation in test readings).
The cost of producing and affixing the label itself is practically insignificant in relation to total direct costs.

b) Indirect costs are mainly a matter of the R & D and investment in which manufacturers may be involved, in order to
make sure that their appliances have an energy efficiency
comparable with the products of their best-placed competitors.
This industrial, commercial and even advertising investment,
could prove to be profitable; its advisability will depend
on the firm's policy decisions and can be judged only by
results. It would be arbitrary to try to isolate indirect
costs under this heading from a firm's general expenditure
on technological research, sales promotion and education
campaigns designed to inform consumers about the performances
of the appliances offered them.

e) <u>Financing the labelling system</u>

85. The cost of launching an energy-labelling scheme is relatively
high, for it has to be sustained over a number of years. Although
there is a basic difference, in this respect, between voluntary and
compulsory labelling, in both cases - although to differing degrees -
the financing of the scheme calls for the co-operation of all involved,
in other words all who have to operate it, or who stand to benefit
if it succeeds or pay the cost if it fails. Manufacturers are involved
because their objective is to secure a high position in the service-
to-consumers league table. The authorities are induced to support a
scheme which is expected to benefit the national economy. Even in
the case of voluntary labelling, the authorities have a decisive
part to play. This includes the provision or supervision of performance-testing, specification and verification structures, a major
role in general information and public education campaigns and the
co-ordination of all measures taken, in close liaison with both sides
of the market. Lastly, the contribution that has to be made by the
trade and consumer organisations, though materially far smaller, is
nevertheless an appreciable factor in the success of the scheme.

f) <u>Procedures for assigning labels, supervision and imposing penalties</u>

86. All the specifications relating to these procedures normally
come under the same authority, but that does not mean that the operations themselves cannot be entrusted to separate organisations,
some of them possibly private. In France, for example, labels are

issued by the AFEI which is a private, non-profit-making association whose Board of Directors is made up of producers' and consumers' representatives in equal proportions.

87. Manufacturers need to be informed, by public procedure, of the general conditions under which they can submit their applications and of the individual specifications relating to each category of product. They must be given the opportunity to submit their comment and their energy-labelling applications within a reasonable period of time and in accordance with the conditions laid down. This procedure may consist (as in the United States) in entering specifications in the Federal Register or (as in France) in subjecting a draft standard to the public enquiry procedure for a period of one month, at the conclusion of which the text becomes final. The draft outline standard, laying down the principles and conditions of application of the system for indicating the energy consumption of domestic appliances, defines the standard's field of application and states what it should contain, the terms on which manufacturers may participate and the arrangements for label supervision, review and withdrawal. The specifications for a particular class of appliance have to indicate its type, the way in which its energy performance is calculated, the tolerances accepted in describing the latter (1) and the terms on which manufacturers may participate in the scheme. Under the voluntary labelling system any manufacturer can withdraw his product from the scheme provided prior notice is given. Manufacturers can also be expelled from the scheme by decision of the public supervisory department or body, if actual performance is found not to tally with the data given. In France, any manufacturer indicating a consumption figure on a label that is proved to be incorrect when the appliance is tested is liable to the penalties provided for by the Act on misleading advertising. If energy-labelling should be introduced in Belgium, the same rules would apply. In Switzerland, the agreements concluded between the private organisations of consumers and producers are put into operation by a joint co-ordinating commission. The manufacturing or distributing firms are responsible for the uniformity, the completeness and the exactitude of the labels. Controls are effected by random selection of models and in the case of errors the co-ordinating commission formulates a special request to the organisations concerned.

---

1) In France, for instance, the energy-labelling rules regarding refrigerators require that the results of a sufficiently large number of tests must not show a deviation exceeding, on average, $\pm$ 5 per cent of the figure declared by the manufacturer.

g) Publicity for the labelling system - information and education
   of the public

88. The publicity given to the labelling scheme, its brand image,
and the steps taken to inform and educate both the trade itself and
consumers regarding its purpose and the advantages to be expected
from it constitute one of the vital conditions for its success. Every
participant in the scheme has to play his own role in making it known,
but all the resources used need to be co-ordinated, sustained and
controlled. In all its diversity, this publicity must have its own
unity and its effectiveness will depend on this unity and on its
continuity. This is a matter for the government. The various operators
involved in such publicity are :

   i) the public departments and bodies responsible for intro-
      ducing the standards and the labelling system, and the
      private associations which collaborate in bringing it into
      wider use. For this purpose, a number of media are available
      to them such as :

      - the publication of an annual report, via official channels,
        giving the list of manufacturers participating in the
        scheme, stating the models to which the labelling system
        is applied and reporting on the findings obtained. In the
        United States, the publication of this report is the res-
        ponsibility of the Department of Commerce;
      - the issue of folders and explanatory leaflets distributed
        on a wide basis and available to purchasers at exhibitions
        and at points of sale;
      - the organisation of debates and demonstrations in the
        press and on radio and television;
      - the organisation of press conferences and the issuing of
        reports of interviews and of handouts to the press in its
        various forms;
      - the possible leadership of a "national campaign" bringing
        together manufacturers and traders, consumers, advertisers,
        advertising agencies and advertising media, with the object
        of publicising the advantages of the labelling system for
        the country's economy. For a relatively short period (15
        days or one month), the campaign might well enjoy the
        voluntary support of a large number of publications and
        other publicity media. "National campaigns" of this type
        have taken place, in France in particular, on subjects of
        general interest (road safety, for example) with undoubted
        success.

ii) Advertisers, traders, advertising agents and public relations firms involved in the sales promotion of the appliances covered by the labelling scheme. Energy-labelling provides them with a new promotional theme and an occasion for opening up a new avenue in research and innovation.

iii) Consumer organisations and institutes, and comparative testing organisations. To the extent that they are involved in drafting standards and in the labelling scheme it is up to them to help to inform the public and ensure that it understands the value of the scheme and how it works. Another way in which they can help is to improve the clarity and presentation of the label, by collecting comments and proposals from users, so that it is readily intelligible to all. What is more, by including energy-performance testing in their facilities, these organisations can help those who read the results of these tests to form a sounder opinion of the quality/price ratio of competing appliances.

iv) Teaching and training staff at the various levels of general and technical education, and all persons in charge of continuing education facilities, have a major role to play in drawing attention to the present reality of energy conservation problems. This instruction should begin from the earliest age and continue throughout school years, the methods used being continuously adapted to the nature of the audience. Use should be made of the literature and teaching aids provided by the relevant technical or educational bodies.

89. National energy-labelling schemes must be compatible with the liberalisation of international trade and with international agreements in conformity with the Trade Pledge (the 30th May 1974 Declaration of the OECD Council meeting at Ministerial Level, renewed in May 1975 and June 1976). Conditions for the assignment of energy labels to domestic appliances need therefore to be discussed and co-ordinated by the Member countries beforehand so as to prevent the possible creation of any obstacles to international trade.

90. The pooling of Member countries' experience would enable them to learn lessons which would save research and wasted time and, possibly, mistakes and could help them to select the most suitable system from among the various schemes in use. This exchange of information could also help them gradually to agree on a number of common guiding principles, particularly as regards the definition of standardized performance-testing methods, the way in which test results are conveyed to the public and certain general rules for assigning labels and supervising the system. This co-operation should in no

way interfere with the original and individual character of each Member country's scheme. Its object would be to arrive, via the necessary diversity of such arrangements, at agreement on points on which their joint success might depend. It could be imagined, for example, that the governments of Member countries which have instituted an energy-labelling scheme for domestic appliances might invite their national standards organisations to submit to the international standards organisations to which they belong (ISO and CEI) the problem of harmonizing the standardized performance-testing methods used at national level. For certain other aspects of the quest for a concensus, other approaches could be considered.

6. OTHER POSSIBLE ASPECTS OF CONSUMER INFORMATION REGARDING ENERGY CONSERVATION

91. Additional measures could be taken in the following fields :

   a) The extension of energy-labelling to other consumer durables, particularly in the fields of heating, and possibly to collective heating installations and lighting.

   b) Heat insulation in buildings. This offers scope for very substantial energy savings in both old housing and new buildings. Stricter thermal insulation and heat control rules can be enforced for new buildings. (1) Furthermore, it is highly desirable that people living in old buildings should be fully informed of the ways in which they can improve the insulation of their homes and of the big savings that this can make. Public information campaigns in various countries, particularly the United States, have had very good results. Advice and government assistance are other ways of prompting investment in improved thermal insulation by private persons.

   c) Among the various measures whose object is to ensure that more national use is made of energy, a noteworthy example is the devising of a standard method for testing motor vehicle fuel consumption on which all published consumption data has to be based. Regulations have been introduced along these lines in several countries including France, Germany, Japan and the United States. More generally, governments might also consider the possibility of setting up a public "innovations" office responsible for supporting R & D on such innovations as might help to reduce household energy consumption.

1) In France, new rules on thermal insulation for new housing were laid down by Interministerial Order dated 10th April, 1974.

IV

SUMMARY AND CONCLUSIONS

92. Consumer information by means of labelling on the product is
an important element of consumer policy and, according to its various
aims and fields of application, either encouraged on a voluntary
basis or prescribed by regulation for certain types of products
by the competent authorities in most Member countries. However,
though labelling is an established instrument of consumer policy,
there is little concrete experience available with regard to the
possible use of labelling in order to achieve economies in energy
consumption.

93. Confronted with the energy crisis and the subsequent problems
of higher energy costs, Member countries have increasingly recognised
the importance of a comprehensive energy conservation policy, which
obviously has to include domestic energy consumption. Measures aimed
directly at the private consumer of energy have therefore been adopted
in the majority of Member countries.

94. The conservation measures directed towards the domestic sector
have mainly consisted of general publicity campaigns attempting
to create more energy consciousness among consumers. Conservation
measures could be very stringent, prescribing for instance a maximum
enery consumption for appliances which could mean banning certain
types of appliances. Compared to this, the energy label can be considered as a more liberal device which is aimed at avoiding waste
of energy without reducing the efficiency of the appliances.

95. However, the energy label is not receiving the same degree of
attention in all Member countries. As at mid-1976, in France, Switzerland and the United States, a voluntary labelling scheme is in
operation where a mandatory system is on the way. It is at the
planning stage in four other countries, Canada, Germany, Japan and
the Netherlands. The Council of the European Communities has issued
a Recommendation on this subject, as set out in paragraphs 59 and 60.

96. There are, indeed, a number of technical and economic problems
to be solved. There is first the technical question of whether it

is possible to measure the energy efficiency of appliances by means of repeatable tests, which implies the definition of an adequate methodology. There is also the economic question whether differences in energy consumption between appliances of a given category are sufficient to justify their measurement and their indication on a label, and, if so, whether there is enough scope for individual savings of energy to make a significant impact on total energy consumption. In connection with these questions, it appears that for a certain number of major household appliances, the technical problems could be overcome and, as regards the possible scope for conservation, the label could be of significance for overall energy consumption.

97. Apart from these technical and economic issues, there exists a number of complex and interrelated problems which, broadly speaking, concern the reactions of the parties involved. There is the basic decision by governments whether the label should be compulsory or voluntary; there is the problem of enlisting the support of consumers, producers and traders and there is finally the practical problem of finding the best way of conveying the information on the energy label to the consumer.

98. If the label is to have any significant impact on buying habits and, in the long run, on product design the label must be applied to a large number of competing products so that the information conveyed on the label reaches a high proportion of potential buyers. In order to be effective, the label would have to oover 80-90 per cent of the market of a particular type of appliance. The question of the attainment of this objective seems to be more important than the question of whether energy labelling should be compulsory or not. It will be up to Member countries to evaluate the possible impact and to decide in the light of their past experience with other types of voluntary, systematic or compulsory labelling schemes whether they consider the energy label in their country capable of reaching and influencing a significant proportion of consumers. In any case, the effectiveness of the label depends essentially on an adequate and continuous supporting publicity campaign and on consumer education.

99. All these problems seem to call for an increased exchange of information on experience gained and on new ideas developed in this field in Member countries, between both those which are actively engaged in operating a scheme and those which are considering the feasibility of introducing one.

100. In addition, when taking regulatory or other action for the implementation of the energy label Member countries should bear in mind their obligations to avoid the creation of non-tariff barriers to international trade.

101. The Committee therefore suggests that :

a) Responsible authorities which are at present developing or operating an energy-labelling scheme should pay attention to the need of harmonizing their action and in particular, to develop common performance-testing methods, and should be ready to consult each other in case of need.

b) Responsible authorities which have not yet undertaken any action in this field should proceed to a study concerning the conditions under which such a label could be operated in their countries.

102. Depending on the results of such studies and the number of countries interested, the Committee could consider whether and in which field it would be useful to develop common guidelines, bearing in mind the need for international harmonization.

Annex I

THE ENERGY LABEL IN CANADA, FRANCE, SWITZERLAND AND THE UNITED STATES
ADMINISTRATION, STRUCTURE, PRODUCTS COVERED

Country	Competent Authorities	Organisation responsible for the allocation of the label	Nature of the label (mandatory-voluntary)	Nature of the labelling scheme (Energy data only - other indications)	Products covered
Canada	Dept. of Consumer and Corporate Affairs	Dept. of Consumer and Corporate Affairs	Not yet decided	Not yet decided	Air-conditioners - refrigerators - freezers - washing-machines - dishwashers - cookers and ranges
France	Energy Conservation Agency to the Ministry of Industry and Research (Agence pour les économies d'énergie du Ministère de l'industrie et de la recherche)	French Association for Information and Labelling (APEI) Association française pour l'étiquetage d'information	Voluntary	Together with other indications	Washing-machines - dishwashers - refrigerators-freezers - electric ranges - gas ranges - electric storage heaters - oil heaters - gas water heaters - gas storage heaters
Switzerland	Consumer organisations and associations of manufacturers of electric household appliances	Co-ordinating Committee, jointly run by consumer organisations and manufacturers	Voluntary	Together with other indications	All electrical household appliances.
United States	The Federal Energy Administration (FEA) under PL 94-163 of December 1975	Federal Energy Administration has primary responsibility for implementation together with the National Bureau of Standards, which will develop the test methods, and the Federal Trade Commission which will develop the labels	Mandatory programme now replaces voluntary programme	Annual energy and cost data	Section 322 (a) A consumer product is a covered product if it is one of the following types (or is designed to perform a function which is the principal function of any of the following types : 1. Refrigerators and refrigerator-freezers 2. Freezers 3. Dishwashers 4. Clothes dryers 5. Water heaters 6. Room air conditioners 7. Home heating equipment, not including furnaces 8. Television sets 9. Kitchen ranges and ovens 10. Clothes washers 11. Humidifiers and dehumidifiers 12. Central air conditioners 13. Furnaces 14. Any other types of consumer product which the Administrator classifies as a covered product under subsection (b) Section 322 (b) 1. The Administrator may classify a type of consumer product as a covered product if he determines that -- A. Classifying products of such type as covered products is necessary or appropriate to carry out the purposes of this Act and B. Average annual pre-household energy use by products of such type is likely to exceed 100 kilowatt-hours (or its BTU equivalent) per year. Listing of consumer products now includes automobiles.

43

EXAMPLES OF ENERGY LABELS APPLIED OR PROPOSED IN FRANCE, SWITZERLAND AND THE UNITED STATES

FRANCE - ENERGY LABEL FOR A FREEZER

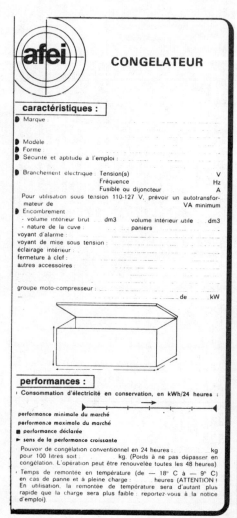

afei

CONGELATEUR

caractéristiques :

▶ Marque :

▶ Modèle
▶ Forme
▶ Sécurité et aptitude à l'emploi :

▶ Branchement électrique : Tension(s)   V
                            Fréquence   Hz
                            Fusible ou disjoncteur   A
  Pour utilisation sous tension 110-127 V, prévoir un autotransformateur de   VA minimum
▶ Encombrement
  - volume intérieur brut ... dm3   volume intérieur utile ... dm3
  - nature de la cuve :   paniers
  voyant d'alarme :
  voyant de mise sous tension :
  éclairage intérieur :
  fermeture à clef :
  autres accessoires :

  groupe moto-compresseur :
                                       ... de ... kW

performances :

▸ Consommation d'électricité en conservation, en kWh/24 heures :

performance minimale du marché
performance maximale du marché
■ performance déclarée
▶ sens de la performance croissante

Pouvoir de congélation conventionnel en 24 heures : ... kg
pour 100 litres soit : ... kg. (Poids à ne pas dépasser en congélation. L'opération peut être renouvelée toutes les 48 heures)
▸ Temps de remontée en température (de — 18° C à — 9° C) en cas de panne et à pleine charge : ... heures (ATTENTION ! En utilisation, la remontée de température sera d'autant plus rapide que la charge sera plus faible : reportez-vous à la notice d'emploi).

conseils d'achat et d'utilisation :

● Tous les appareils sont construits pour une tension nominale de 220 volts.

● Assurez-vous que la prise de courant est raccordée à la terre et que la tension du réseau électrique est appropriée à l'appareil.

● Cet appareil pèse à vide ... kg. N'oubliez pas pour choisir son emplacement, que chargé il peut être 3 à 5 fois plus lourd.

● En l'absence de méthode de mesure, le niveau sonore de l'appareil en fonctionnement relève de la perception et de l'appréciation personnelle. En cas de bruits anormaux, référez-vous à la notice d'emploi et, éventuellement, faites appel à votre installateur.

● Toute denrée congelée ayant subi un début de décongélation par suite d'un arrêt de l'appareil (interruption de courant, panne etc.) ne devra pas être recongelée mais consommée dans le plus court délai.

● Pour le bon usage de votre appareil, en particulier son chargement, son entretien et son utilisation économique, et pour obtenir les meilleurs résultats en congélation, référez-vous à la notice du constructeur et reportez-vous, s'il y a lieu, à une brochure spécialisée sur la congélation.

garantie et service après-vente :

● Pour la garantie du constructeur, reportez-vous au certificat de garantie qui n'est valable que revêtu du cachet du vendeur et daté.

● Si ce dernier vous propose une garantie et des services supplémentaires, faites préciser (sur la facture par exemple) la nature des prestations.

   ↦ de mise en service et de vérification de l'état de bon fonctionnement.

   ↦ assurées dans le cadre, ou en plus, de la garantie du constructeur et les restrictions correspondantes.

   ↦ assurées après expiration de la période de garantie.

● Renseignez-vous sur la Norme Expérimentale X 50-001, contrat de service de bon fonctionnement à la vente et après la vente d'un appareil d'équipement ménager.

délai de réclamation de l'acheteur :

En cas de contestation des mentions portées sur l'étiquette tout acheteur peut envoyer à l'A.F.E.I. une réclamation dans un délai de 3 mois à partir de la date d'achat qui doit figurer sur le bon de garantie.

Association Française pour
l'Etiquetage d'Information
18, rue Saint-Marc
75002 Paris
T. 233.89.12

afei

dossier N°
Tous droits de reproduction réservés.

INFORMATION DES CONSOMMATEURS
Déclaration de marchandise
Machine à laver ▬▬▬
Modèle pour fixation

1. Données générales
Origine: Suisse
Garantie: 1 an
Service assuré par: ▬▬▬▬
Directives d'installation: français/allemand/italien
Mode d'emploi: français/allemand/italien

2. Données pour l'installation
Tensions de raccordement/fusible: 3×380 V $+$ N $+$ T, 1×380 V $+$ N $+$ T
                                   commutable /10 A
Puissance absorbée: 5,1 resp. 3,9 kW
Genre du raccordement électrique: avec câble et fiche
Raccordement pour l'eau: avec tuyau blindé avec raccord $3/4''$
                                   ou tube de cuivre
Installation d'écoulement: avec tuyau d'écoulement $\varnothing$ 19/28 mm

3. Caractéristiques de l'appareil
Matériau pour l'extérieur: en tôle galvanisée revêtue par
                                   poudrage électrostatique, recouvrement:
                                   en acier au chrome-nickel 18/8
Matériau pour l'intérieur: cuve, tambour, récipient produits de lessive,
                                   porte, pompe en acier au chrome-nickel 18/8
Orifice de remplissage: devant $\varnothing$ 27 cm
Capacité: 3,7 kg de linge sec (12 l/kg)
Cotes extérieures: 61/57/90 cm (larg./prof./haut.)
Poids: 105 kg
Programmes de lavage: 11 programmes principaux
                                   4 programmes partiels
Durée du programme avec
dégrossissage et cuisson à 95°C: env. 90 min.
 — consommation horaire
     d'électricité: env. 3,5 kWh
 — consommation horaire d'eau: env. 120 l

4. Accessoires
 — compris dans le prix: 1,5 m tuyau métallique blindé
                                   avec raccord $3/4''$,
                                   2 m tuyau d'écoulement $\varnothing$ 19/28 mm
                                   avec coude 180°, 1,7 m de câble 5-phas.
 — non compris: fiche, montage et installation

La conception et la présentation de cette déclaration de marchandise ont été
fixées en collaboration avec la communauté de travail des organisations pour
la protection des consommateurs FSC/FPC.

ASDF Corp.   Model 5508A10

8,000 Btu per hour
(cooling capacity)

115 volts   860 watts   7.5 amperes

Data on this label
for this unit certified by

energy guide

EER=9.3

Energy Efficiency Ratio expressed in Btu per watt-hour

For available 7,500 to 8,500 Btu per hour 115 volt
window models the EER range is

EER 5.4 to EER 9.9

For information on cost of operation and selection of correct cooling capacity, ask your dealer for NBS Publication LC 1053 or write
to National Bureau of Standards, 411.00, Washington, D.C. 20234

IMPORTANT...

for units with the same cooling
capacity, higher EER means:
Lower energy consumption
Lower cost to use!

Tested in accordance with

U.S. DEPARTMENT OF COMMERCE • LABELING PROGRAM
• ENERGY CONSERVATION

Energy Guide

Data on this label for this unit certified by

ASDF Corp Model 77A
16.0 Cubic Foot Automatic Defrost
Combination Refrigerator-Freezer

Cost of Energy
$6.30 per month

This cost is based on use under standard test conditions and an electric rate of 4¢ per kilowatt-hour (kWh).

The cost of energy will vary with how you use your unit and with your electric rate. For tips on saving energy ask your dealer for NBS Publication LC 1055 or write to National Bureau of Standards. 441 00. Washington, D.C. 20234.

To estimate your cost at your local rate use the table below.

If your electric rate per kWh is	Your monthly cost of energy will be approximately
2¢	$ 3.20
4¢	$ 6.30
6¢	$ 9.50
8¢	$12.60
10¢	$15.60

Comparison Information

The ranges of cost of energy for all brands of 14.5 to 17.5 cubic foot refrigerator-freezers with various defrost systems for which information is available are given below.

Type of Defrost	Approximate Cost of Energy per month at a rate of 4¢ per kWh
Automatic	$4.20 to $7.20
Partial Automatic	$3.50 to $5.90

Energy Consumption
158 kilowatt-hours per month

Under standard test conditions.

OECD SALES AGENTS
DEPOSITAIRES DES PUBLICATIONS DE L'OCDE

ARGENTINA – ARGENTINE
Carlos Hirsch S.R.L.,
Florida 165, BUENOS-AIRES.
☎ 33-1787-2391 Y 30-7122
AUSTRALIA – AUSTRALIE
International B.C.N. Library Suppliers Pty Ltd.,
161 Sturt St., South MELBOURNE, Vic. 3205.
☎ 69.7601
658 Pittwater Road, BROOKVALE NSW 2100.
☎ 938 2267
AUSTRIA – AUTRICHE
Gerold and Co., Graben 31, WIEN I. ☎ 52.22.35
BELGIUM – BELGIQUE
Librairie des Sciences
Coudenberg 76-78, B 1000 BRUXELLES 1.
☎ 512-05-60
BRAZIL – BRESIL
Mestre Jou S.A., Rua Guaipá 518,
Caixa Postal 24090, 05089 SAO PAULO 10.
☎ 216-1920
Rua Senador Dantas 19 s/205-6, RIO DE
JANEIRO GB. ☎ 232-07. 32
CANADA
Publishing Centre/Centre d'édition
Supply and Services Canada/Approvisionnement
et Services Canada
270 Albert Street, OTTAWA K1A OS9, Ontario
☎ (613)992-9738
DENMARK – DANEMARK
Munksgaards Boghandel
Nørregade 6, 1165 KØBENHAVN K.
☎ (01) 12 69 70
FINLAND – FINLANDE
Akateeminen Kirjakauppa
Keskuskatu 1, 00100 HELSINKI 10. ☎ 625.901
FRANCE
Bureau des Publications de l'OCDE
2 rue André-Pascal, 75775 PARIS CEDEX 16.
☎ 524.81.67
Principaux correspondants :
13602 AIX-EN-PROVENCE : Librairie de
l'Université. ☎ 26.18.08
38000 GRENOBLE : B. Arthaud. ☎ 87.25.11
GERMANY – ALLEMAGNE
Verlag Weltarchiv G.m.b.H.
D 2000 HAMBURG 36, Neuer Jungfernstieg 21
☎ 040-35-62-500
GREECE – GRECE
Librairie Kauffmann, 28 rue du Stade,
ATHENES 132. ☎ 322.21.60
HONG-KONG
Government Information Services,
Sales of Publications Office,
1A Garden Road,
☎ H-252281-4
ICELAND – ISLANDE
Snaebjörn Jónsson and Co., h.f.,
Hafnarstræti 4 and 9, P.O.B. 1131,
REYKJAVIK. ☎ 13133/14281/11936
INDIA – INDE
Oxford Book and Stationery Co.:
NEW DELHI, Scindia House. ☎ 47388
CALCUTTA, 17 Park Street. ☎ 24083
IRELAND – IRLANDE
Eason and Son, 40 Lower O'Connell Street,
P.O.B. 42, DUBLIN 1. ☎ 74 39 35
ISRAEL
Emanuel Brown :
35 Allenby Road, TEL AVIV. ☎ 51049/54082
also at :
9, Shlomzion Hamalka Street, JERUSALEM.
☎ 234807
48 Nahlath Benjamin Street, TEL AVIV.
☎ 53276
ITALY – ITALIE
Libreria Commissionaria Sansoni :
Via Lamarmora 45, 50121 FIRENZE. ☎ 579751
Via Bartolini 29, 20155 MILANO. ☎ 365083
Sous-dépositaires:
Editrice e Libreria Herder,
Piazza Montecitorio 120, 00186 ROMA.
☎ 674628
Libreria Hoepli, Via Hoepli 5, 20121 MILANO.
☎ 865446
Libreria Lattes, Via Garibaldi 3, 10122 TORINO.
☎ 519274
La diffusione delle edizioni OCDE è inoltre assicu-
rata dalle migliori librerie nelle città più importanti.

JAPAN – JAPON
OECD Publications Centre,
Akasaka Park Building,
2-3-4 Akasaka,
Minato-ku
TOKYO 107. ☎ 586-2016
KOREA – COREE
Pan Korea Book Corporation
P.O.Box n° 101 Kwangwhamun, SEOUL
☎ 72-7369
LEBANON – LIBAN
Documenta Scientifica/Redico
Edison Building, Bliss Street,
P.O.Box 5641, BEIRUT. ☎ 354429 – 344425
THE NETHERLANDS – PAYS-BAS
W.P. Van Stockum
Buitenhof 36, DEN HAAG. ☎ 070-65.68.08
NEW ZEALAND – NOUVELLE-ZELANDE
The Publications Manager,
Government Printing Office,
WELLINGTON: Mulgrave Street (Private Bag),
World Trade Centre, Cubacade, Cuba Street,
Rutherford House, Lambton Quay ☎ 737-320
AUCKLAND: Rutland Street (P.O.Box 5344)
☎ 32.919
CHRISTCHURCH: 130 Oxford Tce, (Private Bag)
☎ 50.331
HAMILTON: Barton Street (P.O.Box 857)
☎ 80.103
DUNEDIN: T & G Building, Princes Street
(P.O.Box 1104), ☎ 78.294
NORWAY – NORVEGE
Johan Grundt Tanums Bokhandel,
Karl Johansgate 41/43, OSLO 1. ☎ 02-332980
PAKISTAN
Mirza Book Agency, 65 Shahrah Quaid-E-Azam,
LAHORE 3. ☎ 66839
PHILIPPINES
R.M. Garcia Publishing House,
903 Quezon Blvd. Ext., QUEZON CITY,
P.O. Box 1860 – MANILA. ☎ 99.98.47
PORTUGAL
Livraria Portugal,
Rua do Carmo 70-74. LISBOA 2. ☎ 360582/3
SPAIN – ESPAGNE
Libreria Mundi Prensa
Castelló 37, MADRID-1. ☎ 275.46.55
Libreria Bastinos
Pelayo, 52, BARCELONA 1. ☎ 222.06.00
SWEDEN – SUEDE
Fritzes Kungl. Hovbokhandel,
Fredsgatan 2, 11152 STOCKHOLM 16.
☎ 08/23 89 00
SWITZERLAND – SUISSE
Librairie Payot, 6 rue Grenus, 1211 GENEVE 11.
☎ 022-31.89.50
TAIWAN
Books and Scientific Supplies Services, Ltd.
P.O.B. 83, TAIPEI.
TURKEY – TURQUIE
Librairie Hachette,
469 Istiklal Caddesi,
Beyoglu, ISTANBUL, ☎ 44.94.70
et 14 E Ziya Gökalp Caddesi
ANKARA. ☎ 12.10.80
UNITED KINGDOM – ROYAUME-UNI
H.M. Stationery Office, P.O.B. 569, LONDON
SE1 9 NH, ☎ 01-928-6977, Ext. 410
or
49 High Holborn
LONDON WC1V 6HB (personal callers)
Branches at: EDINBURGH, BIRMINGHAM,
BRISTOL, MANCHESTER, CARDIFF,
BELFAST.
UNITED STATES OF AMERICA
OECD Publications Center, Suite 1207,
1750 Pennsylvania Ave, N.W.
WASHINGTON, D.C. 20006. ☎ (202)298-8755
VENEZUELA
Libreria del Este, Avda. F. Miranda 52,
Edificio Galipán, Aptdo. 60 337, CARACAS 106.
☎ 32 23 01/33 26 04/33 24 73
YUGOSLAVIA – YOUGOSLAVIE
Jugoslovenska Knjiga, Terazije 27, P.O.B. 36,
BEOGRAD. ☎ 621-992

Les commandes provenant de pays où l'OCDE n'a pas encore désigné de dépositaire peuvent être adressées à :
OCDE, Bureau des Publications, 2 rue André-Pascal, 75775 Paris CEDEX 16
Orders and inquiries from countries where sales agents have not yet been appointed may be sent to
OECD, Publications Office, 2 rue André-Pascal, 75775 Paris CEDEX 16

OECD PUBLICATIONS, 2, rue André-Pascal, 75775 Paris Cedex 16 - No. 37.983 1976

PRINTED IN FRANCE